Canalside Walks in the Pennines

Spurbook Footpath Guides

The Ridgeway Path
Walks in the Chiltern Hills
Afoot in the Yorkshire Dales
Walks along the South Downs Way
Afoot in Hertfordshire
Afoot in Surrey
Walks along Offa's Dyke
Walks along the Ridgeway
Walks in Avon
Walks in Berkshire
Walks in Buckinghamshire
Walks in the Cotswolds
Walks in Devon
Walks in Exmoor
Walks in Hampshire
Walks in the Hills of Kent
Walks in the Lake District
Walks in Oxfordshire
Walks in the Peak District
Walks in the Surrey Hills
Walks in Sussex
Walks in the Yorkshire Dales
Surrey Rambles
Waterside Walks in West London
The Pennine Way
Canalside Walks in the Pennines

Canalside Walks in the Pennines

A.J. Pierce

Frederick Warne

Published by
Frederick Warne (Publishers) Ltd
40 Bedford Square
London WC1B 3HE

Cover photo: Simon Warner

ISBN 0 7232 3064 1

Printed and bound in Great Britain by
Galava Printing Company Ltd, Nelson, Lancs.

Contents

Page

Introduction 7

The Leeds and Liverpool Canal

Walk 1 Leeds to Shipley - 12¾ miles (20.5 km) 9
Walk 2 Shipley to Keighley - 5½ miles (9 km) 13
Walk 3 Skipton to Gargrave - 4½ miles (7 km) 16
Walk 4 Gargrave to Barnoldswick - 7 miles (11 km) 19
Walk 5 Barnoldswick to Burnley - 11 miles (17.5 km) 22
Walk 6 Burnley Town Walk - 3½ miles (5.5 km) 25

The Huddersfield Broad Canal and The Calder and Hebble Navigation

Walk 7 Huddersfield to Cooper Bridge or Brighouse - 3¼ miles or 5¾ miles (5 or 9 km) 28

The Huddersfield Narrow Canal

Walk 8 Huddersfield to Marsden, Standedge Tunnel - East End - 7¾ miles (12.5 km) 32
Walk 9 Diggle to Stalybridge - 8 miles (12.5 km) 35

The Peak Forest and Ashton Canals

Walk 10 Marple Top Lock to Whaley Bridge - 6½ miles (10.5 km) 39
Walk 11 Dukinfield Junction to Marple Top Lock - 7 miles (11 km) 43
Walk 12 Dukinfield Junction to Ducie Street - 6¼ miles (10 km) 46

The Rochdale Canal

Walk 13 Sowerby Bridge to Todmorden - 10 miles (16 km) 49
Walk 14 Todmorden to Littleborough - 5½ miles (9 km) 54

Walk 15 Littleborough to Rochdale - 4½ miles (7 km) 56
Walk 16 Rochdale to Manchester - 12 miles (19 km) 59

Further Reading 62

Note: All distances are approximate

Introduction

This book contains sixteen walks along the towing paths of six canals in the Pennine area each route having been walked by the author in the recent past.

The task of dividing the many miles of towpath in this area into walks ranging from family strolls, to a serious effort, was not easy. The choice is entirely my own and the walks have been chosen for many reasons.

These walks are therefore a personal choice and a personal viewpoint, though where facts are stated I have made every effort to verify their accuracy.

The walks covered range from West to East and the scenery passed through varies from heavy industry to the splendid grandeur of the Pennine Hills. Each walk has much to recommend it, historically, architecturally or scenically, in most cases all three, and taken together the walks form a unique way of exploring some of the finest hill scenery in the country without actually walking up a steep hill!

At the time of writing the Huddersfield Narrow Canal and the Rochdale Canal except for a short length in Manchester, are unnavigable and can only be appreciated on foot or in parts by a lightweight transportable boat.

All the towpaths covered in this book are in at worst reasonable and at best very good condition but all form ideal walking ground, though it must be remembered that towpaths can be very muddy.

The walks have been chosen so as to be lengthened or shortened at the discretion of the walker and they do not follow consecutively from each other; they are written as I saw them.

I hope you gain as much enjoyment as I have from these walks and meet as many helpful people to whom I offer my sincere thanks.

Notes

1) With the exception of the Rochdale Canal all the canals in this book are under the jurisdiction of the British

Waterways Board. Permission to walk the towpaths of these canals is not required but in the main they are not public rights of way and consequently members of the public walk them at their own risk. On certain lengths local authorities have established rights of way over the towpaths.

On the Rochdale Canal public pedestrian rights of way have been granted over most lengths, including the whole of those under the West Yorkshire County Council and in the Rochdale Metropolitan District and the remaining lengths are the subject of current discussion. In the past the company has charged 20p to cover the position of towpath walkers.

2) On all these walks the use of public transport is recommended wherever possible. Returning to your starting point having left your car there, can be difficult on some walks. On some of the walks I have left my car at the end of a walk and used public transport to arrive at the beginning. Making your own arrangements with the help of a good map and local transport timetables is the best recommendation I can make. I have not included bus route numbers to avoid confusion if route number changes have been made. I must point out that I have have found all bus companies concerned to be most helpful.

The Leeds and Liverpool Canal

WALK 1

Leeds to Shipley

12¾ miles (20.5 km)

THE LEEDS AND LIVERPOOL CANAL: The canal was the first trans-Pennine canal to be started and the last to be fully opened, being constructed under Acts of 1720 to 1819. It is 127 miles (204 km) in length with a summit level of 487 ft (148m) above sea level with a total of 91 locks of which 44 are on the Leeds side of the summit. It also has numerous swing bridges.

Access to canal: Below Leeds City Station there is room for car parking. Cross the River Aire on the A639 road bridge and the junction of the canal with the river will be seen. Turn right alongside the former canal company warehouse and follow this road past a disused canalside crane. The canal office is immediately in front of you. Cross the lock bridge and take the towpath, turning west.

The walk begins in the heart of Leeds, the canal office building being worthy of attention, particularly its carved inscription. The towpath here is wide and firm. The railway which is to accompany us for much of the journey is above and to the right at this point, but soon crosses the canal. The River Aire is also a constant companion and as you pass under the railway arches it appears below the canal level and to the right.

As you leave Leeds a number of fine bridges should be noted, particularly the railway bridge denoted as 225F. The first lock is a single chamber 62ft (19m) in length by 14ft (4m) wide. Soon the first of the two-rise locks, Oddy lock, is reached. A feature of this canal is the use of riser or staircase locks where the top gates of one lock rise from the bottom gates of the next, a fascinating arrangement. Spring Gardens lock, reached next, provides a good illustration of the simple paddle gearing on the lock gates, the rising and lowering of

N
A639
A64
M1
LEEDS
M621
A647
A6120
A657
SHIPLEY
BINGLEY
A650
KEIGHLEY

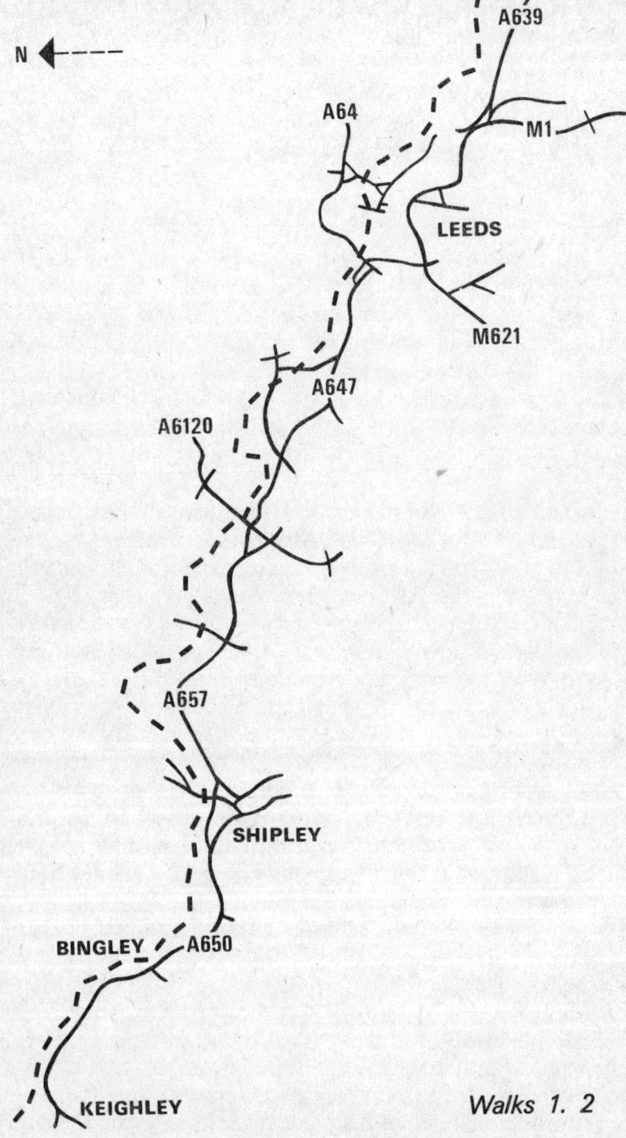

Walks 1. 2

which by a windlass admits water into and out of the lock.

Bridge 225B looks as if two guarding statues should be in residence. Industry still surrounds you at this point, but do not forget that canals and industry were built to complement each other, though almost without exception industry now turns its back on the canals.

By canal road bridge 225A a former working boat can be seen at its last moorings beneath the decaying remains of the covered loading bay of a canalside warehouse. The width of the canal here gives an indication of the amount of traffic once loading and unloading at this point.

Soon more open countryside is reached beyond Armley Mill Bridge. Looking carefully at this bridge, (225) you will note that on the eastern abutment a roller is still in place to protect the towropes used in horse-drawn days from chafing the bridge stonework. Grooves cut by countless ropes can be seen here and also at the western end. Many bridges once had such rollers but most have now disappeared.

To the right Kirkstall power station dominates the scene. Kirkstall was once coal-fed by the canal with its own canal arm. The lines of mooring posts along this arm bear witness to past trade, as do the sunken remains of two boats.

The smell of Kirkstall Brewery pervades the air combining with views of the twelfth-century Cistercian Kirkstall Abbey to the right, together with its folk museum. This is well worth a visit. The single-rise Kirkstall lock is soon reached followed by the Kirkstall Forge lock, the first three-rise lock to be encountered. The name of this lock relates to the industry prevalent here at one time.

A delightful wooded stretch abounding with wildlife brings you to Newley or Newlay lock, again a three-rise, and soon Rodley on the outskirts of Leeds is reached. This is a pleasant spot, the canal curving past a public house, aptly but not originally named *The Barge*, and a small but thriving boatyard. Rodley too provides a good example of a swing bridge.

Now out in open countryside and accompanied by the railway and river the canal swings towards Apperley Bridge. Here a riser of two locks lifts the canal to the side of the BWB maintenance workshops. These buildings, and the cottages just below the locks, have been canal property almost from its opening. In these workshops many lock gates have been made and masons, blacksmiths and carpenters have played their role in the canal's history. Today the sanitary station and other facilities for the pleasure boater are housed in former stables.

Leaving Apperley Bridge the canal begins a huge curve around the hillside through which the railway is tunnelled. The scenery here is pleasant enough except for the huge sewage complex close by the canal. More swing bridges cross the canal, but do not hinder the walker and quite soon the Field three-rise lock lifts the canal another 25ft (7.5m) towards the summit. The canal continues through a deeply wooded section clinging to the hillside above the River Aire and in spring bluebells abound in this area, as do anglers all the year. Shortly after bridge 211 the railway rejoins the canal and from here into Shipley industrialization creeps back along the canal banks.

As bridge 208, a fine swing bridge, is approached you will notice a now truncated branch of the canal to the left. This was the site of the former Bradford Canal now closed and filled in over most of its length. The canal architecture at this point and in the canalside cottages a little further on is well worth noting as is the 1911 Leeds and Liverpool Canal Company notice. The footbridge crossing the canal here is hauntingly named, Gallows Footbridge.

The walk ends at the new concrete bridge 207C carrying the main road over the canal. Turn left here and join the main Leeds Road, where both bus and rail services to Leeds are available.

WALK 2

Shipley to Keighley

5½ miles (9 km)

Access to canal: Bridge 207C on Otley Road. Rail and bus services close by. Car parking is difficult but can be found. On arrival at the towpath turn west.

Immediately on reaching the towpath one is surrounded by canalside buildings, the most impressive of which is the warehouse to your left. This was finely built to last and while it stored a wide variety of goods, wool was the predominant product transshipped here. A superb Salvation Army Citadel can be seen behind the warehouse. A short way out of Shipley the canal and towpath are separated by railings, supposedly erected to prevent horses falling into the water at places where traffic was heavy. To the right an interesting group of stone-built cottages stands incongruously with the new Inland Revenue offices. Tales of the uses of these cottages vary somewhat, but certainly the date 1790 over No. 6 appears to be authentic.

The tale is told that part of the buildings was once a hostelry for boatmen and that stables still exist under cottage No. 9, though I have not seen them. Tales are also told of a murder at these cottages and of them belonging to local quarry owners. Whatever the truth they are a reminder of former canal days.

You will soon reach Saltaire, the canal itself dominated by the mills of Sir Titus Salt. The town is well worthy of exploration, easily made from bridge 207A. The whole community was built for Salt and planned by him to provide his employees with their every need, except for a public house. Today the estate of some 850 houses, church, library, school and almshouses, still thrives and is a unique monument to one man's contribution to the Industrial Revolution.

Returning to the canal, the parkland laid out by Salt enhances the River Aire while the canal advances to Hirst lock, a single-rise closely followed by another swing bridge. At the lockside a well-worn bollard, narrowed by the constant friction of ropes, stands as yet another reminder of a bygone age and is pleasing to the eye. To the right a garden centre may cause you to divert for a time.

From here the canal becomes wide and clear, as it passes through a delightfully wooded length with the River Aire to the right. Returning again to legend, certain boatmen believed this length to be haunted and no boatmen would moor for the night here. Today this length, culminating in the seven-arch Dowley Gap aqueduct, is truly delightful in any season. Look out for the stone distance marker.

Dowley Gap aqueduct, carrying the canal over the River Aire, is a low-arched structure, particularly pleasing to the eye when looked at from the river bank below, which is possible from the western end. The next bridge is a roving or changeline bridge, carrying the towpath to the other side of the canal. The design of bridges such as this enabled the horse to change sides without the towrope being unhitched. Dowley Gap two-rise locks are delightfully situated.

Above the locks at bridge 205 the *Fisherman's Arms* provides refreshment. From here into Bingley the canal is pleasant enough despite a nearby sewage works but Bingley itself closes in with industrial buildings. The canal is in fact almost in the centre of the town, a convenient gap in the car park wall providing access. Opposite the car park are the mills, which until quite recently received all the coal for their boilers by canal. Bingley Station is also close at hand. Under the road bridge and around the bend you find Bingley three-rise locks accompanied by a recently cleaned mill dated 1871 whose engine-house wall carries an interesting carving. Having lifted the canal nearly 30ft (9m) the next pound (the distance between two locks) brings you to one of the wonders of the waterways, the Bingley five-rise lock. Opened in 1774 this staircase lifts the canal some 60ft (18m) and you will no doubt be glad you do not have to operate the paddles and gates. The five-rise always attracts spectators. At the top of the locks a plaque at the towpath side recalls the celebration of the bicentenary in 1974 and a walk over the swing bridge to the lock cottage will reveal a plaque in the wall giving details of the original engineer. Opposite this cottage former canalside stabling is again in use for modern boating purposes.

From here to Gargrave the canal is lock-free and the walk

from Bingley to Keighley is most pleasant and scenically rewarding as the Aire Valley unfolds. The canal is now high above the river and a number of swing bridges bring you to the Granby swing bridge. Here the *Marquis of Granby* provides refreshment while a short distance away East Riddlesden Hall built in the seventeenth century, can be visited, as it is open to the public by the National Trust.

The next swing bridge is Stockbridge or Bar Lane swing bridge and is in fact the nearest point to Keighley reached by the canal. In this area can be seen extensive former canal warehousing of different periods from red brick to local stone. Indeed on the present BWB office can still be deciphered the remains of the former wording of the Leeds and Liverpool Canal Company. Turn left into Bar Lane from the canal, and examine the warehouses. Then join the main A650 road for a bus back to Shipley.

WALK 3

Skipton to Gargrave

4½ miles (7 km)

Access to canal: Car parking is available close by the Springs Branch of the canal, situated left off the A65 into Coach Street, just outside the town centre. By bus, walk out of the bus station, turn left on the A629, then immediately left into Swadford Street to the canal bridge, turning west on to the towpath. By rail, cross the A59 directly on to the canal.

It is worth exploring the canal environment in Skipton. The junction with the Springs Branch, or Lord Thanet's Canal, is now marked by a hire boat base and the branch itself is worth exploration. Only half a mile (0.8 km) in length it was built to carry limestone from Lord Thanet's quarries at the castle in Skipton to the main line of the canal. The canal boats were loaded via a wooden chute from a railway high above and the remains of the chute can be found behind the castle today, though this is a shorter chute than the original, as stone descending from a great height caused much damage to the boats. The branch certainly makes a pleasant walk.

Back at the main line, the remains of Skipton's former glory as a canal centre can be seen. At one point Skipton saw much activity and was the headquarters of one of the largest private carrying fleets on the canal. The canal company itself also had a depot here and when the canal opened in 1773 the first two boatloads of coal were sold in Skipton at half the previous price, which brought great jubilation to the town.

Today Skipton still has a canal atmosphere, an old warehouse converted to a public house and other former canal buildings as an outdoor pursuits shop. Trips on the canal by boat can also be arranged from here.

Having absorbed all this, the walk begins accompanied by mills, the canal crossing a most unusual little aqueduct, very easy to miss, and there are huge numbers of ducks and geese and a few swans. The inevitable swing bridge soon appears and equally quickly open countryside is visible. Leaving

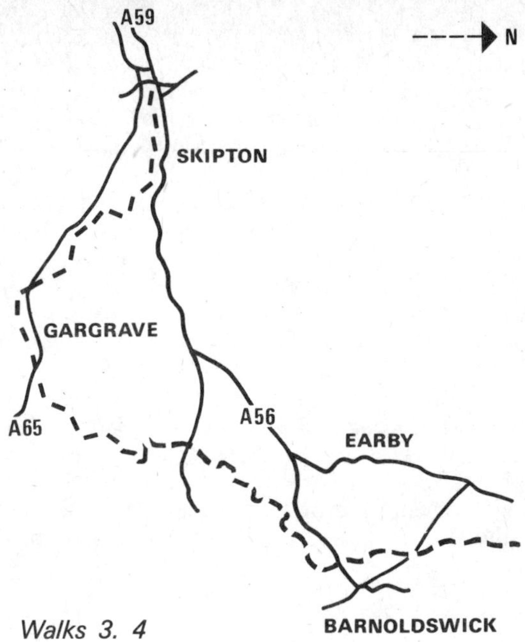

Walks 3. 4

Skipton the canal is now crossed by a new bypass but this is soon left behind.

The water here is wide, deep and clear, a feature in fact of almost all this canal. At bridge 175, Niffany Swing Bridge, there are two items of note. One is the excellent house standing on the inside of the canal bend; the other is the fact that for a short distance the towpath becomes part of the road before re-establishing itself through a wooden gate. This next length of towpath can be particularly difficult to walk in summer due to the profusion of undergrowth, but the effort is well worthwhile as a variety of plants can be found. Willowherbs and harebells predominate.

From here to Gargrave the views on both sides of the canal are truly magnificent, the head of the Aire Valley on one hand, the rising limestone hills of the Yorkshire Dales on the other. This country may give you some indication of the difficulties faced by the canal builders who began this length in 1791.

The canal is again interrupted by a bypass crossing but is soon left to its tranquillity, being interrupted only by a couple of swing bridges. Holme Bridge (172A), a rather functional bridge not of the original canal era, carries the A65 over the canal and close by a farm provides overnight caravan accommodation. Under the bridge, Holme Bridge lock heralds the end of the 16-mile (26-km) lock-free pound as it lifts the canal further towards the summit. The small village of Gargrave is now visible, well worth exploring before returning to Skipton. At the centre of the village a bridge over the River Aire divides the village in two. The church is mainly Victorian but with a sixteenth-century tower. The village is stone built and pleasant to walk in.

However, before exploring the village, continue along the canal past Eshton Road lock to Higherland Bridge. Turn left here into the village and a return bus to Skipton.

Though a short walk, this length of canal has much to offer in the way of splendid scenery and gives time to admire both Skipton and Gargrave with their respective attractions.

WALK 4

Gargrave to Barnoldswick

7 miles (11 km)

Access to canal: By bus or rail to Gargrave. Car parking in Gargrave. Bus timetables should be consulted for the return journey. Cars may also be parked at Barnoldswick and the bus service used to Gargrave; this involves changing at Skipton, and this walk could be combined with Walk 3.

The canal is located at Higherland Bridge, just north of the A65. Turn west along the towpath.

This walk will take you through spectacular changing scenery. The walk begins at Higherland lock, the third lock of the six-lock flight. Locally this lock is known as Highland lock and several members of a famous local canal family were born here. Turning left, (west) along the well-maintained towpath, you find yourself under an arch of leaves in summer. Anchor Bridge is closely followed by Anchor lock, lifting the canal a further 9ft (3m), and on the opposite bank the garden of the *Anchor Inn* is a fascinating place in itself. Anchor lock is a fine example of the simplest type of ground paddle gear. Here the lever is simply moved to uncover a hole in the wall of the canal to allow water into the lock.

Next is Scarland lock closely followed by the curiously named Stegneck Bridge and lock. Passing under the railway bridge you find yourself almost immediately crossing the River Aire on the Priest Holme aqueduct. A short clamber down to river level will be well rewarded. It is also worth looking at the infant Aire and comparing it with the volume of water flowing through Leeds.

The canal has now almost reached its most northerly point and soon turns south towards Bank Newton locks, but before these are reached admire the beautiful countryside through which the canal is passing. At bridge 168 the towpath changes sides and becomes part of the road for a short distance before returning to its rightful place through a gate. Ahead is the Bank Newton flight of six locks, one of the most picturesque on the canal. A boat hire firm has taken over the former carpenters' yard. Each lock has a beautiful, simple footbridge at its tail, again a feature of this canal. It is amazing that although the canal locks are wide, deep and long, so much is strikingly simple. Certainly the surroundings of this hire base would encourage even the most dubious of hirers. At lock 40, the fifth one of the flight looks back over the way you have walked. Surely this must be one of the finest views anywhere in the Dales.

The canal now enters a rather wooded twisting section, and at bridge 165, the Newton Changeline Bridge, the towpath again joins the road for a while. Rejoining the canal itself you are on an incredible pound or pool, as the distances between locks are known on this canal. It is one of the most unusual stretches of canal I have ever walked, at times almost rivalling parts of the Oxford Canal. Horses may be banned from the towpath but cows are obviously not. A television mast serves as a landmark here. It approaches you from all sides leaving you wondering where it will next appear. The curves of the canal are made all the more interesting if a boat is travelling at the same time and you can follow its progress.

This is one of the remotest stretches of the canal with real Pennine scenery, few trees and only sheep for company. Civilization begins to return at East Marton with its most unusual bridge. This is known as Double Arch Bridge and the reason will be only too obvious. There are interesting tales about this bridge, including one of a cement lorry which went over the bridge into the canal. *The Cross Keys Inn* is above the canal to the right on the A59. The wood-lined cutting, through which you have been passing and which the bridge spans, is soon left and the canal returns to rugged moorland country. Enjoy this length as it cleaves its way through the hills. Greenberfield locks are the next point of real canal interest. Just before the locks, the towpath again changes sides over a temporary bridge. Greenberfield locks again form a fine flight though all are single. Alongside the top lock is the lock-keeper's cottage and a small building dating from the 1890s through which the supply of water from Winterburn reservoir reaches the canal. By the bottom lock you will have noticed a now infilled arm of the canal. This was once the original line of the canal which reached the summit level by means of a three-rise staircase lock. These rises were replaced in 1820 by three single locks. As you walk up the flight you will notice the depression where the canal ran and also a canal bridge on the former line. You are now on the summit, 487ft (148m) above sea level, and you will no doubt have noted the pride the lock keeper has in his flight.

From here to Barnoldswick the canal scenery is less dramatic and more urban as it approaches the town of Barnoldswick. By leaving the canal at bridge 153 refreshment and transport home can be obtained.

An extra short distance can be added to the left just beyond bridge 153, where the now-hidden Rain Hall Rock Branch lies. No sign of it is seen from the towpath as the first stretch is infilled but beyond that a watered canal with two tunnels and an aqueduct can be found. The branch was originally built to serve a quarry.

WALK 5

Barnoldswick to Burnley

11 miles (17.5 km)

Access to canal: Bus or car to Barnoldswick and access to the canal at bridge 151 can be gained from the B6385. Check times of public transport for your return journey if necessary. Alternatively the canal can be rejoined at bridge 153 at the end of Walk 4. Turn west along the towpath.

Leaving Barnoldswick behind, open views are soon reached by Salterforth with the *Anchor Inn* providing refreshment. The towpath along here has a variety of wild flowers and plants. The remains of the once busy wharf at Salterforth can be discerned. The former Lancashire-Yorkshire Boundary is crossed just before Mill Hill Bridge. At Foulridge a diversion to the walk has to be made. However, before this diversion, former stables, warehouses and the wharf can be seen and you can peer into the 1640-yard (1500-m) tunnel which has no towpath, hence the diversion. This tunnel caused serious problems for the engineer, Robert Whitworth, and delayed the opening of the canal as a through route to Burnley. It took five years to build, eventually opening in 1796. The passage of boats through the tunnel was never easy, vessels originally being legged through, men lying on their backs and 'walking' on the tunnel walls or roof. Leggers were replaced in the 1880s by steam tugs which were later superseded by the diesel engine. Those who clung to horse power were given a tow through the tunnel.

Foulridge's main claim to fame concerns the cow which was reputed to have swum through the tunnel on 24 September 1912 from west to east, to be revived by alcohol. A photograph of the cow hangs in *The Hole in the Wall* public house. It is locally believed that this in fact happened twice. The tale is also told that two men died when diesel power was introduced to the tunnel, being overcome by fumes. There is little danger of this for the walker as you follow the horse route over the tunnel.

Returning from the tunnel mouth to the wharf, walk up the street towards *The Hole in the Wall* pub in the village centre,

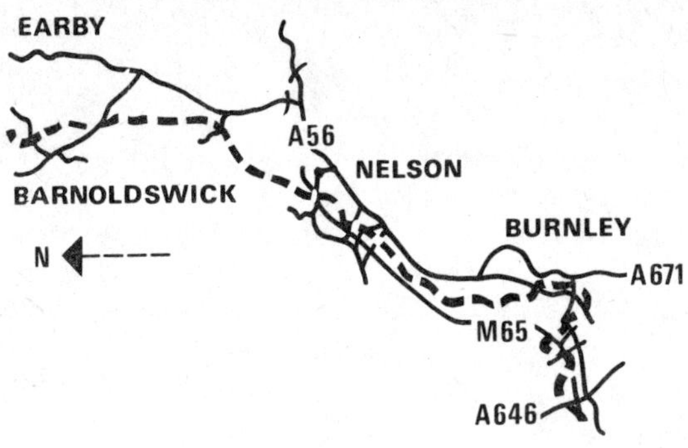

Walks 5. 6

calling to see the photograph of the cow. Turn right here and follow the Salterforth sign until you reach a bend with a bus shelter facing you. Turn left past some attractive cottages until about half a mile (0.8 km) ahead a gate complete with BWB signs can be found. By now you will no doubt have spotted the tops of the tunnel ventilation shafts denoting the line of the tunnel which is actually not too far below ground level. From this gate the horse path heads for the western portal, accompanied now by the disused Colne to Skipton railway line. You will find yourself at the head of the portal with horse path sloping down to water level. From here look across to the other side of the canal where the leggers' stone hut can be seen. Note the stone steps and handrail leading from the canal level. As you walk back on to the towpath note the large number of mooring rings set in the wall, a testimony to the queues of boats that once waited to pass through the tunnel. You now find yourself in a cutting, indeed the tunnel itself is really a covered-in cutting.

This is a pleasant stretch of open countryside and soon the seven Barrowford locks come into view marking the end of the summit pound and descent into Lancashire. Alongside the locks is Barrowford Reservoir, a feeder for the canal and also storing excess water from the summit. These locks are exceedingly well kept and lower the canal a total of 69ft (30m). Under the road bridge and just above the second lock is the site of the former wharf serving nearby Colne. A large warehouse stood here until recently but today only the old toll office remains, though this is still in use. A careful look above the window will reveal a point of interest.

Passing these last two locks the environment changes from rural to industrial and Colne Water is crossed by the three-arched Swinden aqueduct. To the right, away from the industry, Pendle Hill dominates the scenery. At bridge 142, Swinden Changeline Bridge, the towpath changes sides as the canal enters Nelson.

Nelson warehouse is a forlorn sight though it was one of the last to be built and indeed has signs of the railway era about it. Through Brierfield the scenery continues as industrial on the left, hills to the right, until the walk ends at Lodge Bridge, No. 134. From here a short walk to the left takes you to a bus stop with a good service to Burnley. Alternatively, you can continue into Burnley itself on Walk 6.

WALK 6

Burnley Town Walk

3½ miles (5.5 km)

Access to canal: Access can be gained at bridge 134, Lodge Bridge, from Barden Lane in the suburb of Reedley. A bus from Burnley town centre brings you close to this bridge. Roadside car parking is available.

This is a lovely walk illustrating fine canal engineering, industrial archaeology and a panoramic view of Burnley. The towpath is in excellent condition throughout and much of Burnley's history can be learned along the way. At Lodge Bridge are two cotton mills: the red brick Barden Mill, still occupied by a cotton manufacturing firm, and Lodge Mill. This is, however, not the original mill, which was destroyed by fire in 1905. The bridge itself, apart from the metal footbridge, is an original canal bridge. Under the bridge is a new housing estate to the right and beyond that views towards Pendle Hill and beyond. The towpath here has much to offer the naturalist with a profusion of plants and some pleasant woodland.

Passing under bridge 134, Heald Bridge, and looking to the left you will find the remains of Reedley Colliery. The canal widens here and an idea can be gauged of the traffic which once used these wharves. From this point it is worth looking over Burnley's rooftops into the distance.

Walking on, the Oswald Street gasholder which has appeared from several angles as the canal has twisted and turned, appears again but the former railway siding complex has now gone. Before reaching the gasholder you will have noted the Colne to Preston railway crossing the canal, and at the sharp right turn two mills built in the 1880s.

The canal now straightens again and it is pleasant to note some landscaping at the rear of one of the mills. New Hall Bridge (132) is next reached and close by was once the site of a boatyard. This stretch of canal is interesting for its variety of mill architecture, particularly on the bank opposite the towpath. Under bridge 131A and looking to the far bank again, you will see an attractive garden which stands out among the industry. Along the towpath on this length is found the first of several ramps down into the canal and out again. These are horse pull-outs. On busy lengths of canals, such as this horses frequently slipped into the canal and these places were provided for their rescue.

Under Colne road bridge the canal swings left, skirting what was once Bank Hall Estate. For a time the area just beyond Colne road bridge was the terminus of the navigable section of the upper part of the canal and this became a very busy area. The first Burnley warehouse built by the canal company was on this site. It has now completely disappeared. A short way past the Bank Hall Miners' Club and on the opposite bank was Bank Hall Colliery, now obliterated and landscaped. The only remaining site is that of a short canal arm. The canal now crosses Brun Aqueduct, an interesting structure, and passes through Thompson Park, a delightful length of the walk, tree-shrouded and with plenty of bird- and fish-life. Turning under Godley Bridge (130H) you are brought to another of the wonders of the waterways, the Burnley Embankment, often called the straight mile though it is in fact some 500 yards (457m) short of this. Carrying the canal 60ft (18m) in the air, it gives spectacular views. Originally, in 1796, it was only 46ft (14m) high but was later raised. As you admire the views do not forget that the embankment also contains two aqueducts, the first known as the Culvert Aqueduct, (a contradiction in terms) over Yorkshire Street, one of Burnley's main roads, and the second over the River Calder. Close by the Culvert Aqueduct once stood a ropewalk, and limekilns also existed in this area.

As you near the end of the embankment, just prior to Finsley Gate Bridge, there is an exit to the town centre.

Under this bridge the BWB maintenance yard is found in good condition. Buildings on this site have always been associated with the canal. There was once a swing bridge here, and stables existed, all now converted to workshops but well worth a close look.

Also worth looking at is the Finsley Gate Bridge (130E), a cast-iron bridge resting on rope-marked stone pillars with the remains of an iron roller fitted to stop the towropes chafing.

The walk from here to Gannow tunnel is an industrial archaeologist's paradise. Hugging the canal almost the whole way are mills interspersed with warehouses and wharfage. Finsley Mill is the first on the right and just before Centenary Way the modern road crosses the canal. Excellent views can be had of an open mill chimney base, while under Centenary Way an original horse pull-out is found. At Manchester road bridge (130B) look closely at the towpath edge to find the remains of toll gates equivalent to those on turnpike roads. Immediately round the corner is the toll house itself, at the beginning of the superb Manchester Road wharf. Happily the area is being restored. This was once the centre of Burnley's canal trade in the early nineteenth century and there is much to see, including a wheel set into the towpath in front of the warehouse entrance. Around this was hooked the boat towrope, so the horse on its way to the stables at the rear, could pull the boat into the wharf. Walk under the huge canopy, built to compete with the covered loading offered by the railways, and imagine the scene; a look below the warehouse doors will show you how much trade took place here.

Spend some time exploring here as there is much to see. Further on, cobbles form the towpath and soon you are in the heart of Weavers Triangle the area of the town most connected with the cotton trade. Under bridge 130, Sandygate Bridge, Slaters Terrace is found to the left. Built about 1850, the upper storey housed weavers and the lower storey served as a warehouse. Walking on you will soon cross a new aqueduct over a motorway extension, which is in itself a tribute to modern canal engineering. Round the bend you come to Gannow tunnel; look carefully here for masons' marks in the portal. The tunnel has no towpath and to finish the walk you must retrace your steps to the new Whittlefield Bridge (129) from where a bus can be caught into Burnley, or alternatively you can retrace your steps to Finsley Gate to appreciate the industrial scenery again.

This walk will be made even more enjoyable by reading *Along t'Cut, a Towpath Trail Through Burnley* published by Lancashire County Council. It was compiled by the Burnley Teachers' Centre and gives excellent detail and maps. Available from local bookshops or Burnley Library.

The Huddersfield Broad Canal and The Calder and Hebble Navigation

WALK 7

Huddersfield to Cooper Bridge or Brighouse

3¼ miles or 5¾ miles (5 or 9 km)

THE HUDDERSFIELD BROAD CANAL: The canal was built between 1774 and 1780 under the guidance of a local engineer, Luke Holt. It is often referred to as Sir John Ramsden's canal as he was its promoter. Although it was linked to the Huddersfield Narrow Canal, through traffic was a problem owing to differences in lock sizes, the locks here being 57ft 6ins by 14ft 2ins (17.5 x 4.3m) as on the Calder and Hebble. The canal is fully navigable into the centre of Huddersfield from its junction with the Calder and Hebble.
THE CALDER AND HEBBLE NAVIGATION was engineered by Smeeton and is a combination of river and canal.

Access to canal: Aspley Basin, the start of the canal, is at the side of the A642 Wakefield Road, on the left as approached from the town centre. To the right is the truncated end of the narrow canal. Turn left into St Andrew's Rd, pass the rear of the coal hoppers and turn left, which brings you to the towpath.

As you walk to the towpath, Aspley Basin, a scene of much former activity, is well worth looking at. Here goods were transshipped to and from the Huddersfield Narrow Canal as through passage was limited to a number of small narrow boats, the lock length of the narrow canal being 70ft (21m) (57ft 6in (17.5m) on the broad canal and the widths 7ft (2.2m) and 14ft 2in (4.3m) respectively). Reaching the tow-path you are faced by a decaying but still elegant eighteenth century warehouse.

The remains of a former warehouse, now much reduced in height, come into view, with some pleasantly arched windows. Immediately in front of you is the unique Turn-

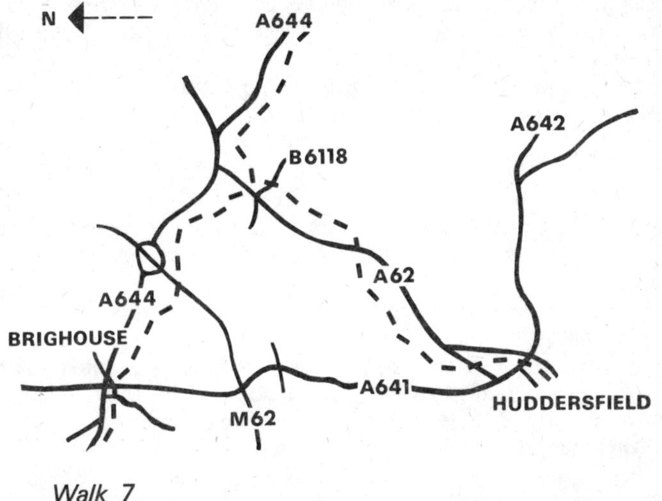

Walk 7

bridge lifting bridge built in 1865. A plaque near the left-hand counterbalance states that this is a locomotive bridge. When raised the entire bridge deck rises parallel with the waterway. The bridge is well worth examining. The area is surrounded by fine stone mills, one dated 1846 with a pleasant stone chimney in front of it.

Walking on past several windlasses on the towpath edge, you can see a slipway on the opposite bank beside a former church or chapel. The remains of an infilled arm can be seen by the side of more early warehouses. The canal now passes under two bridges, the second of which, an arched concrete structure, is similar to the one on the Blackburn flight of locks. Swinging sharply left, then right, you pass under the main A62 road and find the corporation bus depot on your left. At the next bridge you will see a well-worn flight of stone steps leading from the towpath to the road above. The canal now passes an incineration plant where the bank has been extensively sheet-piled. Passing on, the canal becomes more rural and to the right Huddersfield Town Football Club comes into view. Red Doles lock is soon reached, with its typical stone bridge at the tail. Former stables can be seen to the left of the lock. The canal is now distinctly more rural and the next lock is soon reached in a pleasant area. Look back over your shoulder at this point for a good view of Huddersfield. As you pass each of the locks on this canal you will find something of note; for example at the next lock, No. 6, an interesting and very simple addition has been made to the balance beams to assist in opening the lock gates. Along this length there are extensive playing fields to the right and in the hills beyond the beautifully-shaped Emley Moor TV mast can be seen. As the main A62 Leeds to Huddersfield road is reached, below lock 5 another fine warehouse can be seen.

The canal soon becomes industrialized again, with extensive works to the left. Below the next lock and to the right on the towpath, a former mileage post, minus its number plate, can be seen. The next interesting point is by the Yorkshire Water Authority Works. A stone bridge support stands at either side of the canal, the left-hand one bearing the inscription 'Private Bridge 1860'. The next bridge has a delightful lopsided look about it, due to the slope of the road down the hill. Five stop planks, to be inserted under the bridgehole in times of a leak, are to be found here. As you emerge under the bridge the next lock, No. 2 is found, and rounding the curve brings you to the entrance lock from the River Calder. This is Cooper Bridge, where perhaps the less energetic could end the walk.

The Calder and Hebble Navigation

Turning left by the bridge over the canal the path towards Brighouse continues up the lane towards the works of Walter Fernley. However, a walk on to the main road bridge over the River Calder is worthwhile here, as Wharf Works is a good example of functional waterway architecture. Having returned to the riverside path you can see the junction of the river and a canalized section. Here are examples of the new BWB road with signs pointing to Sowerby Bridge, Wakefield and the Huddersfield Broad Canal and to the danger of a weir. Towards Sowerby Bridge the towpath is now accompanying a river section, which it follows to Kirklees low lock where the navigable channel and towpath swing left, leaving the river.

This area is in pleasant open countryside. Just beyond the lock are two mileposts in the towpath side. Kirklees top lock is next approached; it has an interesting wheel set in the stone edge just before the lock entrance, no doubt used for tow-ropes. The canal now passes through a very rural stretch, wooded on the left-hand side, but sight and sound are dominated by the M62 motorway bridge, a good example of modern bridging. You will notice here a number of Lancashire and Yorkshire railway posts at the side of the towpath. The Anchor pit lock (Kirklees flood gates) is approached next. Once again the river becomes a navigable channel here and these lock gates remain open at all times except in cases of flood when they can be closed off to prevent flooding on the canalized section. From here follow the lane alongside the river, rejoining the riverside path where the lane swings sharp left. Leaving the path at Brighouse is not too easy and you may wish to leave it here instead.

However, if you continue, the path, which is very little used, is interesting. Reaching the centre of Brighouse you will see with some dismay the well-restored entrance lock to the canalized section on the opposite side of the river. A former toll office and other buildings are visible here as well as a former bridge which would have carried you to the other side. However, to leave the river you must scramble up the embankment, make your way left into Dyehouse Lane and turn right again into Birds Royd Lane. Turn right again into Huddersfield Road, from where a bus will return you to Huddersfield. It is also worth walking along the canal, to the right, to view Brighouse Basin.

The Huddersfield Narrow Canal

WALK 8

Huddersfield to Marsden
Standedge Tunnel - East End

7¾ miles (12.5 km)

THE HUDDERSFIELD NARROW CANAL: This canal is just under 20 miles (32 km) long and includes the longest canal tunnel in Britain at Standedge. The tunnel is 3 miles 418 yards (5.2 km) in length and 645ft (197m) above sea level. Unfortunately no towpath exists through it. The canal was begun in 1794 and opened throughout in 1811, the second Pennine waterway to be opened. There are 74 locks - 42 on the Huddersfield side and 32 on the Ashton side. The canal runs between Huddersfield and Ashton-under-Lyne in Lancashire, although it is no longer open to navigation, being officially abandoned in 1944. As it is unnavigable the walker has a unique opportunity to explore it.

Access to canal: Car parking is available in Huddersfield and the town is well served by bus services. The walk begins on the A642 road bridge on the opposite side of Aspley Basin, close by Huddersfield Polytechnic.

Here the remains of a former transshipment warehouse can be seen, where goods could be shipped from the Broad to the Narrow Canal, or vice versa. The canal has been restored and landscaped in front of the Polytechnic as far as the first lock. At the first lock a temporary road causes the canal water to be culverted and the surroundings change dramatically. The chamber is intact to its full length, an unusual sight as most locks on this canal have been at least partially, if not fully, infilled or cascaded. The lock gates still stand but are in very poor condition. Passing lock 2E and under Queen St South, the canal is culverted under a new building and the towpath abruptly ends, but you will have experienced a wild, neglected urban waterway.

Now retrace your steps to the previous bridge with towpath access, turn right, right again, and again right into Chapel

towpath has changed sides and returned again in its passage through Uppermill and the superbly-named Cloggers Knoll bridge. Having no towpath you will have to cross over the bridge.

Walking on from Wade lock, you next reach Halls lock 20W, where grooves in the corner stones of cottages show wear from towropes. The canal has now become far less dramatic than it was towards the summit, and as lock 19 is passed signs of past industrialization come into view. Work has recently been done here, leaving the towpath messy, but this is compensated by the site of an old lock hut on the lockside.

The Royal George Mills are also to be seen on the left here, a fine example of mill architecture. Passing lock 18W you reach the Royal George aqueduct, a pleasant sight after a rather depressing length of canal. You are now 488ft (150m) above sea level and the two-arched aqueduct carries the canal over the River Tame. On the far side a windlass can be seen and there is also a stone on the aqueduct dated 1797. The canal here is again somewhat depressing as far as lock 17W. Close by lock 16W, where the towpath changes sides, the *Tollemache Arms* provides refreshment.

The channel soon improves and by lock 15W the Tame valley improvement scheme is visible, where an attempt has been made to improve the canal environment, including provision of fishing lagoons. The canal along this length twists and turns, following the River Tame below. At the tail of lock 14W the towpath changes sides and the canal begins to enter an industrialized area with Milton Mill dominating the scene. Past lock 13W, Micklehurst Brook floods across the towpath, making life a little difficult. Soon the 205-yard (185-m) Scout Tunnel is approached. Although this has a towpath, both ends are sealed. However, it is easy to walk round, and the walk is pleasant. The section between locks 11W and 10W is heavily wooded, indeed the canal and towpath are in danger of becoming divorced from each other completely, though there is much to interest the naturalist here. Hartshead Power Station now dominates the scene and the site of a former swing bridge is reached. The canal is culverted here for some distance, as pylons carry cables over it. The final length into Stalybridge, though industrial, is attractive and the water is deep and clear. The walk ends with an improvement scheme at Mottram Road. Turn right here to visit the centre of Stalybridge. The canal from Mottram Road is infilled through Stalybridge but its course can, with perseverance, be followed. However, the final length of canal to its junction with the Ashton Canal is best approached from the Ashton Canal end.

The Peak Forest and Ashton Canals

WALK 10

Marple Top Lock to Whaley Bridge

6½ miles (10.5 km)

THE PEAK FOREST AND ASHTON CANALS: The Peak Forest Canal runs from a junction with the Ashton Canal at Dukinfield to former basins at Buxworth, (formerly Bugsworth) and a branch to Whaley Bridge, very much regarded as the main terminus of the canal today. Its main purpose was to carry limestone from the quarries of the Peak District; the engineer was Benjamin Outram. The canal is built on two levels connected by 16 Marple locks. At 518ft (160m) above sea level, the canal from Marple to Whaley Bridge is the highest stretch of navigable water in the canal system. Building began in 1794 and work was complete by 1804.

The Ashton Canal was a slightly earlier venture, being authorized in 1792. The Ashton Canal connects with the Rochdale Canal at Piccadilly, Manchester; with the Peak Forest at Dukinfield; and with the now unnavigable Huddersfield Narrow Canal half a mile (0.75 km) from the Peak Forest Junction.

There were three branches on this canal: Hollinwood, Fairfield and Stockport, all now abandoned, but the Hollinwood branch is traceable. In the late 1960s the Ashton Canal was the subject of an exciting restoration scheme which resulted in reopening it to navigation.

Access to canal: Peak Forest Canal - from the centre of Marple village walk along the flight of locks, joining it where the Stockport-New Mills Road (B6101) crosses the canal. If arriving in Marple by rail, turn right out of the station until the locks are reached. Convenient car parking is available in Marple. If using public transport to return to your car it is advisable to check local timetables.

This is a fascinating scenic walk with a good towpath. The

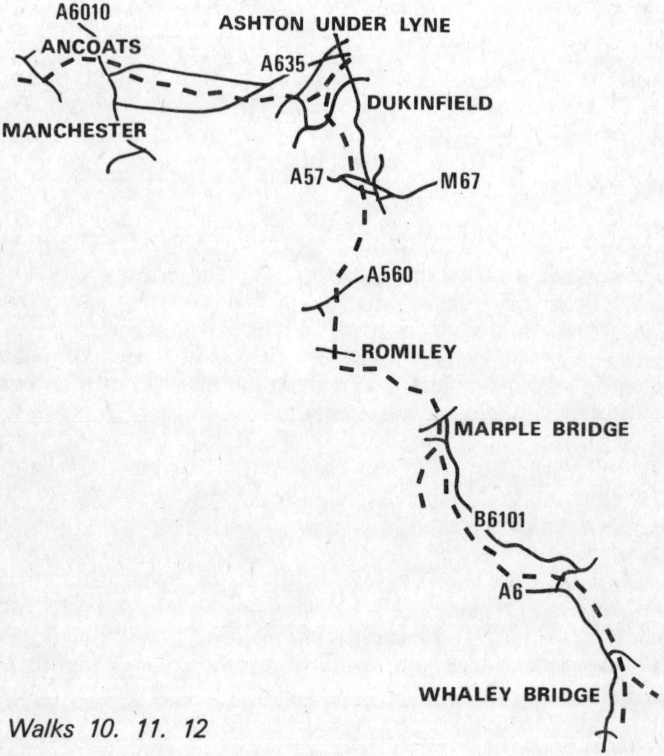

Walks 10. 11. 12

walk begins among buildings of canal interest. The well-kept house opposite the towpath was once the home of a boat-building yard, its dry dock now serving as a sunken garden. To your right there is a fine farmhouse. This is a popular mooring spot for boats. Already to the east, left, you will have noticed the excellent scenery towards the village of Mellor and the hills beyond. As you walk along the towpath look for the house below canal level to the left, which had a long association with Samuel Oldknow, one of the canal's main promoters. At Brick Bridge 19, the towpath changes sides and it quickly becomes clear that this is a contour canal clinging to the valley side high above the River Goyt. Throughout this walk views over the Goyt valley towards the hills in the distance are spectacular and the canal bank itself is well wooded. This is a walk of bridges; stone, swing and even new lift bridges. By bridge 21, Rawton Walls Bridge, two delightful cottage gardens sweep down to the canal. The wonderfully named Turflea Swing Bridge, number 21, is now a lift bridge. Between the next two bridges, 22 and 23, Strines aqueduct is crossed. This can be easily missed but is well worth a close look, being a small, sturdily built stone structure with a delightful curving fascia giving it strength from the base. Beneath the whole structure runs a cobbled road. Looking into the valley from just beyond this point, the village of Strines can be seen with its cotton mill hugging the river. Looking farther up the valley a glimpse of the small town of New Mills can be had but there are other delights first. Bridge 23 is a finely proportioned stone bridge, named Stanley Hall Bridge. This length of the canal is prolific in plant and wildlife and in autumn is a blackberry-picker's paradise. Wood End Bridge, 24, is again a new lift bridge with an accompanying footbridge. The swing bridge, 25, is in need of replacement, but because of its age and condition adds something to the landscape. To the right of the canal here, and approached up the hill from bridge 26, the village of Disley appears; along this length a serious breach of the canal occurred in 1972. You cannot, of course, fail to appreciate the beautifully maintained frontage as the canal bends towards Disley. Between bridges 26 and 27 Bowater's Paper Mill dominates the canal, though the swans that frequent this length compete for dominance. Walking on and looking over the valley, you view the towering Kinder Scout, often known locally as the end of the Pennine Chain, behind New Mills. The canal along this part is pleasant and just before New Mills is reached, at bridge 28, a now largely overgrown winding hole, where the 70-ft (21-m) narrowboats could turn, is seen. New Mills is, as its name implies, a town of mills and these crowd the canal around.

To the right at bridge 28, a thriving new boatyard exists. As you move on into open country, splendid views are again seen to the east, with a fine railway viaduct in the valley bottom. Meanwhile the A6 thunders to the right.

Bridge 29 is a delightful stone structure in a secluded

setting. Furness Vale where extensive new moorings have been created is soon reached and the Manchester-Buxton railway now accompanies the canal. Leaving Furness Vale, a fine new lift bridge is approached and further on at Bridgemont is a swing bridge in a beautiful wooded setting. Bridge 35 leads you over the canal, but spend some time here. Under the canal at this point is a horse tunnel, little known and little used. Walk up to your left towards Bugsworth (Buxworth) where restoration work on the former limestone loading complex is under way. Return to bridge 35 to continue your walk into Whaley Bridge.

The end of the canal is heralded by an 1832 warehouse, now restored. This was the transshipment warehouse between the canal and the Cromford and High Peak Railway running over the hills to the Cromford Canal, a link between two great mill owners, Oldknow and Arkwright.

WALK 11

Dukinfield Junction to Marple Top Lock

7 miles (11 km)

Access to canal: Peak Forest Canal - at the roundabout joining Ashton Road and Cavendish Street in Ashton under Lyne, follow the new ring road towards Manchester for a few hundred yards. Turn left into Welbeck Street, then left into Portland Place, and right into Portland Street South. Car parking is available in this area.

Note: A torch may be useful when negotiating Woodley Tunnel, which is dark and frequently wet.

Leaving Portland basin, at Dukinfield junction the Peak Forest Canal heads south over a fine stone aqueduct over the River Tame. You almost immediately pass under a bridge built in 1845 and rebuilt in 1978. Under the bridge you will find a landscaped picnic area, part of the Tame Valley Improvement Scheme.

Approaching Hyde the canal passes under a new motorway bridge, complete with overflow weir. The scenery so far has been somewhat industrial and this continues through Hyde. Access into Hyde can be made at bridge 6, which also transfers the towpath from the west side to the east. The works seen to your right belong to Joseph Adamson and Company, and dominate the waterway at this point.

At bridge 7, again transferring the towpath to the west bank, the remains of a roller mounting can be seen where once a roller was placed to prevent towropes chafing on the stone. The bridge itself is an attractive stone structure. The canal here, and for much of its journey, is on a steep embankment above the River Tame. The site of a former bridge with stone steps leading nowhere can soon be seen. The tall, solid Gee Cross mill is next approached, Gee Cross today being a small suburb of Hyde. Judging by the interesting cluster of cottages near the mill it was once a small independent community. The canal now shows how much of a contour canal it really is, sweeping and curving its way towards Marple and still on one level. At bridge 9, Windlehurst Bridge, an interesting use has been found for lengths of railway line.

The canal in this area as it approaches Woodley and Bredbury is very rural but by looking a little farther afield, you are constantly aware that you are near the large conurbation of Manchester and Stockport, indeed the tall tower of the CIS building in Manchester can be seen to the west and the constant roar of aircraft above denotes flight paths of Manchester airport. At bridge 11 the remains of the fulcrum for the swing bridge are all that can be seen of the former crossing point. Woodley is soon reached, with an interesting mill to the right. The former mill pond is now in a terrible condition, but it is nevertheless a fine example of a mill with its pond. Woodley is passed through quickly and Woodley Tunnel is found beyond the skewed railway viaduct. The *Navigation Hotel* can be seen to the left and cobbled setts lead to steps to the roadway above. The tunnel has a towpath which is very wet—this is where a torch is an excellent idea. Bridge 13 is a splendid, high-arched bridge with rope marks at both ends. The railway closes in to the left here and soon the canal passes through two railway bridges, connected together but of differing bore. This gives a most peculiar appearance from the end you are approaching.

Romiley is the next village to be reached, with access at bridge 14. Just under this bridge former canalside wharfage can be seen, now well filled in and covered over. A little way on the canal crosses a small but again beautifully built aqueduct, well worth looking at from road level. Becoming very rural and again on an embankment, the canal passes through a most attractive woodland setting, with gardens to match reaching down to the water.

Hyde Bank Tunnel is reached in a deeply wooded setting and with what looks to be very limited headroom. As there is no towpath here, you follow the horse path over the tunnel entrance and under a most attractive little bridge, with an almost sculptured appearance. At the end of this track a wooden signpost directs the way to Marple aqueduct, and where the road splits a small wooden post again points the way with its yellow arrow. Another post appears past the farm and the towpath is soon regained. Further along the embankment, you pass through the now opened-out site of a former tunnel, quite an eerie experience in fact but very interesting to see what remains. Almost immediately the magnificent aqueduct over the River Etherow is reached and is well worth the scramble down towards the river to obtain a clear view. Crossing the aqueduct 100ft (30m) above the river, with the railway viaduct alongside, makes you appreciate early engineering. Under the railway bridge and past former canal buildings to the left, and over the bridge, the first of the 16 Marple locks raising the canal some 209ft (64m) is reached. All the locks have a nearly equal rise of 13ft (4m). This is one of the finest flights of locks in the country, again having undergone recent restoration. Between the locks wide pools ensure sufficient water to fill the deep chambers.

The towpath up the flight is in excellent condition. Each of these locks has something to commend it. Most have a bridge at the tail, often with interesting masons' carving in the keystone. Up the flight beyond lock 8 the main road to Marple Bridge crosses the canal with the next lock immediately into the bridge. To the left is a magnificent warehouse, once belonging to Samuel Oldknow, into which boats could pass for loading and unloading. Further up the flight the Marple Stockport road crosses at the interestingly named Posset Bridge with its horse tunnel. It is reputed that Oldknow offered the workers a posset of ale for completing the job on time. The third arch of this bridge once carried an arm of the canal to the medieval-looking lime kilns. A walk along the road, towards New Mills, will give a good view of these. Continue up the flight to reach the summit.

There is so much more than space permits to say about Marple, Oldknow and the locks. You will find that Marple is a very interesting place, well worth finding out more about.

WALK 12

Dukinfield Junction to Ducie Street

6¼ miles (10 km)

Access to canal: Ashton Canal - as for the previous walk, at the roundabout joining Ashton Road and Cavendish Street in Ashton under Lyne, follow the new ring road towards Manchester for a few hundred yards. Turn left into Welbeck Street, then left into Portland Place, and right into Portland Street South. Car parking is available in this area.

The walk really begins at Dukinfield junction itself where Portland basin, though now a shadow of its former self, is worth spending time examining. Though destroyed by fire some years ago the remains of the fine warehouse still contain three arched entrances by which the canal entered the premises. The cranes used for loading and unloading were worked from the power of a water wheel which still remains, but in very poor condition, in its wheelpit. The magnificent bridge over the canal, reminiscent of bridges on the BCN, now well maintained and restored, blends with this scene of former activity. The aqueduct carrying the Peak Forest Canal is close by and the adjoining landscaped area overlooking the River Tame is the site of what has become the annual Tameside Rally.

Begin by walking along the towpath towards Manchester, westwards. The canal is raised on a high embankment above the River Tame, and surroundings here are very industrial. Along this length an overflow weir carries excess canal water into the river. Soon the railway joins the canal to your left and the remains of a former transshipment dock, long enough for two narrow boats end to end, can be found separated from the canal by the towpath. Guide Bridge Station is passed and nearby, on the opposite bank, at least one householder has shown what can be done with a garden in an industrial landscape. Soon the canal becomes really depressing with the towpath in quite a poor state but persevere for things do improve and a look into the surprisingly clear water will reveal shoals of tiny fish, a most encouraging sign. As you approach Droylsden things improve decidedly with open spaces and a rural atmosphere once the canal has passed a large jam-making factory. Improvements to the towpath have taken place here, and one is reminded at times of the Regents Canal walk in London. Soon the top of the 18 locks is reached, beginning with the Clayton 9 and to the right can be seen the entrance to the former Hollinwood Branch which ran for just over 4½ miles (7.25 km). Though much of the branch is now infilled it is worthy of exploration. The top lock shows all the signs of the restoration work that took place on this canal between 1968 and 1974. A boathouse dated 1833 can be found in this vicinity and a pleasant bridge can also be seen. At one time this lock was one of a pair, symbolic of the heavy traffic use. From here the walk towards Manchester is almost unique as the skyline constantly changes before you. As the locks fall towards the city there is much of interest and the public house beside lock 13 provides a welcome. At this point a swing bridge crosses the lock, adding to the interest. Between locks 10 and 11 you cross a roving bridge under which the former Stockport Branch of the canal left the main line. With careful exploration this branch can be traced.

Lock 9, by a chemical works, is something of a landmark in the history of this canal as it was restored and landscaped before the canal was reopened to navigation.

The Beswick locks numbers 4 to 7 are surrounded by an industrial landscape that is far from oppressive; indeed the disused gasworks at the head of the locks is almost an architectural work of art! Close by, Bradford Colliery had its own branch of the canal between locks 6 and 7. The canal soon crosses the River Medlock and is itself crossed several times by bridges, the towpath changing sides at a particularly interesting bridge. The last flight of locks comprises the three Ancoats locks which are reached by walking between factory walls, as the canal creeps towards the very centre of Manchester. Between the first and second locks, yet another branch, the Islington, ran off to the right. This flight of locks again shows signs of renovation and is well used by local people, particularly at lunchtimes and weekends. From the foot of the locks it is a short walk over the Store Street aqueduct, a strong single-arched structure. Just beyond here road access is gained, where a left turn will take you towards Piccadilly Station which has been clearly visible for some time. By turning left, and then right into Dale Street, the entrance to the Rochdale Canal and its junction with the Ashton can be seen.

This is a real canal enthusiast's walk and provides a back door into or out of Manchester. There is much to see throughout the length of this walk in which the towpath is in excellent condition.

The Rochdale Canal

WALK 13

Sowerby Bridge to Todmorden

10 miles (16 km)

THE ROCHDALE CANAL: The Rochdale Canal, though today unnavigable, is a magnificent waterway. It climbs through 92 locks in its 33-mile (53-km) journey from Manchester to Sowerby Bridge in Yorkshire, where it meets the Calder and Hebble Navigation. It was engineered by William Jessop assisted by William Crossley, although Rennie is frequently credited with its construction. The Act for the construction was passed in 1794 and ten years later it was open throughout, the first trans-Pennine canal to be completely opened. Its locks are large at 74ft by 14ft (23 x 4.25m). The Rochdale was undoubtedly a very well built canal with its locks of fairly uniform rise and eight reservoirs to feed its enormous appetite for water. By 1937 there were no more boats working over the whole length, and except for the section between Piccadilly and the Castleford Junction with the Bridgewater, the canal was abandoned in 1952. Today, still in private hands, part of it forms a linear park, and in other areas restoration is in progress. This is a beautiful canal for walking.

Access to canal: Car parking is available in the centre of Sowerby Bridge. Good bus services exist. A convenient place to begin the walk is at the junction with the Calder and Hebble Navigation, just below the main street of Sowerby Bridge. Head west.

Before beginning the walk it is worth looking from the towpath at Sowerby Bridge basin and its accompanying warehouses, once an important canal centre and now undergoing a revival. Turning west, you find that the site of the first lock is partially obliterated. Cross the main road, Wharf St, and pass behind the *Commercial Inn*, continuing over the car park. The canal here is steeply culverted, where two locks

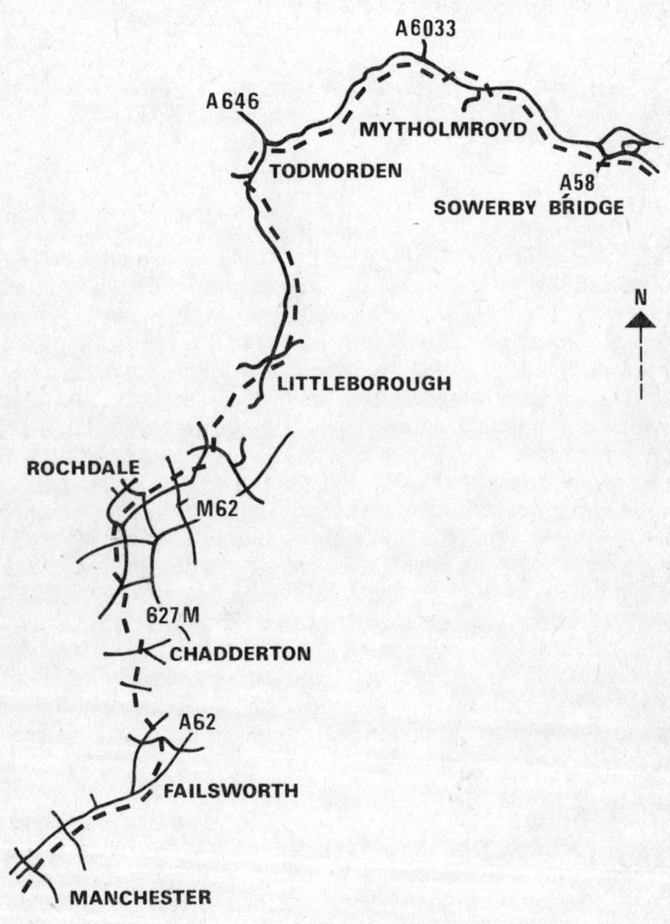

Walks 13. 14. 15. 16

have been replaced. Looking directly ahead you will see the water from the canal, with the towpath on the left, or southern bank. Looking at the grill-covered entrance, and listening to the water, will give you an idea of the gradient involved.

However, once the towpath has been gained you will find fine buildings to your left, clearly once associated with canal carrying. High to the right, above a virtually sheer rock face, a church stands dominating the scene. As you walk along you cannot fail to notice the clear water of the canal. You are virtually walking above the roof tops and weaving sheds with their distinctive U-shaped roofs.

Sowerby Long Tunnel, all 43 yards (40m) of it, is the next item of interest. This is hewn out of solid rock and is the only tunnel on the canal. Despite this, or because of it, the Sowerby portal is, like so much on this canal, well built and impressive. The original length of the tunnel, incidentally, was three yards shorter. The canal now enters a delightfully wooded stretch with a wide towpath, a veritable haven for wildlife of many types. High Royd Bridge follows, and the once-culverted Longbottom Bridge, unusual in that its construction uses gabions, more often seen at the seaside. Beyond this bridge the Calder valley opens out to the left and a profusion of wild plants abounds; in autumn this is excellent blackberry territory. A short way further on a miserable section is experienced where the canal has been ignominiously channelled into a concrete conduit. This has an unkempt look, so uncharacteristic of this canal.

Luddendenfoot is reached after some 2½ miles (4 km) and the canal passes through the park. As you pass below Station Road Bridge a short aqueduct over the River Ludd is encountered, while the council car park is on the site of the former wharf. Beyond the village a straight length of canal follows the Burnley-Halifax Road and the canal here is a pleasure to walk; the number of anglers will testify to its fishing qualities.

Brearly Lower lock begins the climb towards the summit, closely followed by Brearly Upper lock. From here to Brearly Bridge is a somewhat depressing length of canal, the channel having been reduced to little more than a watercourse along the canal bed and the spread of undergrowth is quite phenomenal, a salutary reminder that a canal can very soon be entirely lost. However, almost as compensation, the length from Brearly Bridge is the haunt of much wildlife. Mytholmroyd is reached soon after you pass beneath a former swing bridge. Between Mytholmroyd and Hebden Bridge is possibly the pleasantest stretch of the walk for the naturalist and ornithologist. The only interruption to this length is at the culverted bridge where the Halifax-Burnley road has been improved. Although the canal runs virtually alongside the road for some distance into Hebden Bridge, it is almost unnoticeable from the road.

The approach to Hebden Bridge is most pleasing, the towpath shows signs of maintenance and for a considerable length the park adjoins the towpath. On the opposite bank fine stone buildings dating back to the days of heavy commercial traffic are pleasing to the eye and historically interesting.

At Blackpits lock, almost in the centre of the town, the towpath changes sides as the canal crosses the River Calder on Hebble End Aqueduct. This is a most interesting area, and a good view of the aqueduct can be gained by staying on the original towpath side and walking over the aqueduct, then up a flight of steps. Once you are back on the towpath, Hebden Bridge itself is well worth exploring. This town appears to have been built out of the rock walls of the hills, and some spectacular views can be seen around here.

As you walk along the towpath through Hebden Bridge there is much to show how the canal served the town with a variety of goods from food to cotton. Houses back almost on to the waterway and the canopy of a warehouse still remains. Sadly this remarkable scene ends at Hebble End Bridge with the canal culverted. Blackpits lock is a fine place to see the two sets of recesses in the lock chamber walls. The idea behind this was to install two sets of gates so that when a boat from the Calder and Hebble, 57ft 6in (17.5m) long, passed through, the 74-ft (22.5-m) locks need not be completely filled, but the idea was never put into practice.

Climbing again past locks 10 and 11, Stubbings locks, you can see the fine architecture of this canal with the short wide pound, or pool, between the locks, and also a lock cottage. The scenery now begins to open out towards the hills and from here to Todmorden the canal passes through a variety of phases, from a wide clear channel, to trickles of water along the canal bed. The towpath is punctuated by weirs for overflow into the river, but stepping stones provide welcome crossing points. The canal climbs up the Calder Valley under the impressive Whitelee Arches carrying the railway line. Between locks 12 and 13 the sewage works is passed and a look into the chamber of lock 13 will reveal the former lock tailgate lying in the water. Just above lock 13 cottages face on to the canal and beyond them a bridge leading nowhere, known as Greenwoods Folly.

At Callis Bridge the Pennine Way crosses the canal. At Burntacres Bridge two wooden rollers deeply grooved by tow-ropes, remind us of the extensive horse-drawn traffic. To the left the hamlets of Mankinholes and Lumbutts cling to the hills. Stoodley Bridge too has its wooden rollers in place.

By Woodhouse Bridge stands the impressive 1832 mill of Thomas Binns, Clogmaker. After two further locks Todmorden is reached, and the towpath emerges in the town centre just beyond the bridge, dated 1864, which carries the A646.

WALK 14

Todmorden to Littleborough

5½ miles (9 km)

Access to canal: Access to the canal can be gained from the bridge over the canal on the Rochdale road out of Todmorden, at the point where Walk 13 ends. There are good bus and rail services to Todmorden and car parking is available. Good public transport services exist for the return journey. Turn west along the towpath.

This is a spectacular walk on which the Rochdale Canal climbs over its magnificent summit level and begins the descent on the Lancashire side. Leaving Todmorden, the towpath has changed to the left-hand side and the area around the first lock, immediately under the bridge, has been landscaped though some evidence of neglect is apparent. The canal curves sharply to the left accompanied by buildings which obviously once utilized the canal. To the right, above a high brick-faced embankment, runs the railway. At a warehouse in this length the remains of a fender at the canal edge can be seen, while the lift towers above. As you continue, Dobroyd Castle dominates the hill to your right. Climbing past further locks, each a masterpiece of engineering, you reach Gauxholme where the railway viaduct, with its dramatic style of architecture, broods over the canal. This is a fine sight for transport and engineering enthusiasts. Hereabouts the former lock house, canal buildings and wharf are to be seen, some still in use. Here is an excellent example of how boats could enter a building to load and unload in the dry. Like all the architecture on this canal the warehouse is of solid stone construction.

At Copperoshouse Bridge, the canal is sadly culverted and the towpath, which changes sides here, is reached again by crossing the road. Another series of locks with interesting names, such as Smithyholme, Pinnel and Alma, raises the canal further towards its summit and you will be well aware of the grandeur of the scenery which you are approaching. The village of Walsden closes in on the canal and is an excellent stretch for ducks, geese and even swans. The water is deep, wide and clear. The church and school are reached by a bridge over the canal. Shortly after this bridge a peep through the mill windows to the right will reveal textile machinery hard at work.

Open country is now reached, Pennine scenery at its best with views which never fail to impress. Climb past more locks to reach the East Summit lock 36. You will not have failed to notice, however, that in some of the pools between locks the canal has been reduced to a trickle. The East Summit lock has been the subject of restoration work by members of the Rochdale Canal Society. It is a heartening sight to see the lock gates in place, the bottom gates having been rescued from the lock chamber itself and repaired. The summit pool, though only three-quarters of a mile (1.2 km) long, can tell many a story. A chemical works once straddled the canal here, boats were frozen in during harsh winters and the pool provided a well-earned breathing space for boat crews weary with lock working. Note the mooring rings along this length.

At the West Summit lock a feeder arm enters the canal from nearby Chelburn reservoir high in the hills. The area around this lock has yet to be restored and is in sharp contrast to the Eastern Summit lock, but is nevertheless interesting. The descent to Manchester now begins, dropping some 520ft (160m). On your way down to Littleborough you will begin to see some of the serious obstacles encountered when restoring to a fully navigable channel. A car park over the canal bed at the Fothergill and Harvey Mill is the first real problem. You will have noticed also a portable building over a lock chamber. Despite this, the view towards Littleborough is pleasant, with rural stretches of canal broken by industrial surroundings and a further culverted bridge just before Littleborough itself. The walk ends at Canal Street, just behind the station. A subway under the railway will take you to the centre of the town.

This is a dramatic but easy walk along a good towpath and provides good family walking through excellent scenery.

WALK 15

Littleborough to Rochdale

4½ miles (7 km)

Access to canal: By bus or rail to Littleborough. Car parking available in Littleborough. Under the station underpass, turn right along Canal Street to Ben Healy Bridge. The towpath is on your right. Turn westwards.

The walk begins along a wide, well-used towpath, and on the opposite bank surface water feeders trickle into the canal. Soon the Armour Hess Chemical works dominates the canal, the pump houses at either end of the works giving a good example of the industrial use to which the canal water is put. Ignoring the chemical works for the moment, pause and look behind you to the splendid scenery.

Shortly after the chemical works some fine stone buildings are to be seen on the opposite bank. Flanked now by modern housing, the towpath becomes part of an access road to a canalside mill. Smithy Bridge now culverted, is reached and gives us a clue to a former occupation here. The road which crosses the canal here leads to Hollingworth lake, constructed to supply the canal with water, a function it still fulfils today. At one time, when local workers could not afford to go farther afield, it was known as the 'Weavers Seaport'. It still attracts large numbers of visitors and is worth a visit if you have time.

Bordered again by new housing, fenced off from the canal, the channel becomes wide and clear and soon Little Clegg footbridge, now a fixed structure, is reached; a close look at the stone supports of the bridge will reveal that this was the site of a swing bridge. The canal curving to the right, the remains of the community around Clegg Hall are reached.

The hall itself, now in ruins, is seventeenth century and the watermill has two tailrace openings into the canal. The small row of cottages once belonged to handloom weavers. Clegg Hall Bridge is a delightful structure, being an early stone construction. Look back at the Clegg buildings through the bridge arch, an attractive sight. The canal now enters a rocky cutting and much of what early canal engineering in these parts was about can be seen. The towpath edge is of solid masonry blocks. A little farther on a Lancashire and Yorkshire Railway boundary post can be found to the right of the towpath. It dates from the 1850s when the Rochdale Canal was leased to the railway company.

Bellfield Bridge is reached and just prior to this another LYR post can be found. Immediately before the bridge the canal widens considerably to allow boats passing under the bridge to align themselves correctly. Beyond Bellfield Bridge you could be forgiven for believing that all the splendid moorland scenery has disappeared. Lowfield Mill is to the left and Coppy Footbridge is passed under. You next emerge on top of Firgrove Bridge under which the canal is culverted. The roar of traffic here from the main road contrasts with the peace of the canal. As you return to the towpath the canal runs between two mills, and mooring rings in the path indicate a former canal connection, the rings being set in square stone blocks. Pass under the railway bridge to the site of the former Moss swing bridge, now reduced to a fixed flat deck. Rochdale's ring road is carried over the canal by the concrete Kingsway Bridge, which at least provides navigable headroom. Beyond Moss Road Bridge the scenery becomes industrial, Rochdale's famous mills hemming in the canal. These mills will delight anyone with an interest in such buildings. The architecture of these mills indicates a pride in their ownership, but sadly many now stand empty.

The towpath itself has changed too. The Job Creation Scheme has produced a transformation. The towpath is surfaced and shrubs have been planted. The whole area is pleasant and safe to walk in and as such attracts people to the canal. Lock 49, Moss Upper lock, is a heartening sight and the first since the East Summit to have lock gates in place. The whole pool between this lock and the next has been well landscaped. Along this length can be seen the remains of an infilled canal arm.

At lock 50 the towpath is carried to the opposite side of the canal and this bridge is well worth studying. The scene here is also quite remarkable. This was once the junction of the main canal and the Rochdale branch, the infilled remains of which are to be seen. There are some beautifully restored buildings opposite you. As you continue along the towpath you will find the Job Creation Scheme is still much in evidence. The Lane bridge is next reached and immediately before it Crossfield Terrace, dated 1896, is a pleasing piece of housing architecture. Over the bridge the canal runs in a narrowed open channel and a run-off weir is to be found. The canal enters the Ashfield Valley Flats development and the next two bridges have been culverted to provide access to the estate. An attempt has been made to fit the canal into the development as an amenity for children. At the Queensway Motorway access road you are forced to leave the canal as it enters a long culvert. From here you can return to Rochdale centre, which is well worth while or continue on Walk 16. To return to Rochdale turn right into Queensway. A local bus can be caught by the houses to the left.

WALK 16

Rochdale to Manchester

12 miles (19 km)

Access to canal: Access is gained by the trans-Pennine Industrial Estate, off Queensway. There is a good bus service to the area, and car parking is available close by. Walk to the towpath by turning right from the roundabout after the motorway access roundabout on Queensway. Turn right again into the industrial estate, and a few hundred yards to the left the towpath is gained.

The walk begins on a surfaced towpath with fine old mills to the left and the new industrial estate to the right. Under Bow Street Bridge the original cobbled setts are still in situ. Arrow Mill to the left boasts a tall brick chimney, a sight rapidly disappearing even in Rochdale. Blue Pitts top lock, 51, is reached, a seemingly deep lock with accompanying buildings along the towpath and the former lock tailgates lying in the bottom of the chamber. Rising to street level to cross the canal, you find a good fish and chip shop and sweet shop before returning to the newly restored Blue Pitts middle (52) and lower (53) locks. The restoration work that has gone on here recently has revealed even more of the fine architecture of this canal as for example the flight of steps alongside lock 52. The path now changes to the east bank and heads for the M62 motorway where the former Heywood branch left the main line. This has now been virtually obliterated. Naturally, the canal is culverted under the motorway and you follow a footpath sign to regain the towpath in a fine rural length before Slattocks top lock is reached. A lock house is to be found here together with a water control point. From here the canal continues its descent through a pleasant flight of six locks with short but wide pools and usually a good number of anglers.

Continuing your walk down the flight you will see an impressive piece of modern church architecture to the west between locks 58 and 59. The canal next passes under a quite impressive skewed brick railway bridge. The west bank soon becomes hemmed in by modern housing as you pass locks 60 and 61, where tailgates lie in the chamber. It is the railway, however, that has most interest at this point, with a magnificent iron bridge known as 'The Iron Donger'. It was built by Messrs Radford of Waterloo Foundry, Manchester, and remains in its original condition. The bridge is certainly worthy of better attention than it is receiving at present.

Lock 62, Coney Green lock and 63, Walk Mill, herald the arrival of a large sweeping curve in a simple but splendidly solid aqueduct, the scene some 60 years ago of a serious breach. For the next mile or so the canal itself and its surroundings are somewhat rubbish-filled and depressing, particularly around the 1905-built Malta Mill. Passing under Laurel Bridge the site of the former swing bridge at Firwood is reached.

Lock 64 is reached while the canal environment is still in a depressing state; nevertheless plenty of anglers can be found. Chadderton Power Station now dominates the way ahead and the canal literally passes through it. Two pylons are firmly planted in the canal bed beyond the power station, with the canal culverted beneath them. The next feature of note is Henshaw Lane Bridge, to be found immediately beyond Hollinwood Avenue Bridge just before the great Ferranti factory. Henshaw Lane Bridge is an interesting early structure but unfortunately in some danger of collapse. The end of the Failsworth pool, one of the longest on the Lancashire side of the canal, is reached at lock 65, Failsworth lock. This is actually the end of the true canal, for from here into Manchester the canal has been adapted to fit into a modern development. At Oldham Road Bridge the canal is culverted and you return to street level where you turn right into the new shopping precinct. Head across the car park, diagonally to the right, where the canal is found again in a straight length to the next culverted bridge. Lock 66 is alongside a chemical works and the lock chamber has been infilled to leave a few inches of water at the top, while the lock tail has been cascaded. Between this and the next lock a very attractive low-rise housing development has taken place. At Poplar Road Bridge a large notice proclaims that you are entering the Greater Manchester Council's scheme to improve the canal environment. The canal itself, although retaining its full width, has been reduced to a few inches deep. The towpath is surfaced and the whole walk is pleasant and interesting.

A pleasing diversion can be found at the *New Crown Inn* which faces on to the canal and serves good beer. Back at the towpath, further modern housing is being built to overlook the canal and there is great potential here. It is sad, however, to note how the locks have been treated, being reduced to almost ornamental cascades, though this must be more attractive than a dangerous derelict canal. A little further along the canal a former winding hole, where boats could be turned, can be seen. At the site of lock 77 the walk ends abruptly, the towpath being fenced off. To resume your walk reach road level, turn left, first right into Coleshill Street, past the dye works, right again into Varley Street and on your left the canal towpath is regained.

The next landmark in a somewhat depressing landscape is the *Navigation Hotel* with its attractive sign, a reminder of former days. Despite the 'modernization' of the canal and modern housing, the canal bridges retain the original flavour of the waterways architecture. Again the surfaced towpath ends with a beautiful bridge leading the canal on to Redhill Street, along which horses once towed their boats. At the end of Redhill Street, you will have noticed that the canal has disappeared under Great Ancoats Street, and a walk across this road will reveal the very sad state of the canal. However, just beyond here the canal becomes navigable again for its journey through the centre of Manchester as part of the Cheshire ring. Walk left along Great Ancoats Street, turn right into Ducie Street, over the junction with the Ashton Canal and right again into Dale Street where the Rochdale Canal Co. headquarters can be found. You are now virtually in the centre of Manchester close to Piccadilly.

FURTHER READING

Along t'Cut - A Towpath Trail Through Burnley, R. Frost, Lancashire County Council 1977

The Huddersfield Narrow Canal - A Short History, D. Charlesworth, Huddersfield Canal Society 1980

The Huddersfield Narrow Canal - A Unique Waterway, Huddersfield Canal Society

A New Canal for Huddersfield, Huddersfield Canal Society 1976

Through Stalybridge by Boat, Huddersfield Canal Society

Fill to't top wi' rubble, Huddersfield Canal Society

The Rochdale Canal, Waterway Handbooks Company

Huddersfield Canals Towpath Guide, Ed. D. Charlesworth, Huddersfield Canal Society 1981

Pennine Turnpikes and Canals, C. Whitehead & M. Johnston, Pennine Heritage Network 1982

Canals and Waterways, A.H. Boddy, Greater Manchester Council 1975

Guide To the Calder and Hebble Navigation, Calder Navigation Society 1975

The Rochdale - an Illustrated History, M.T. Greenwood, (M.T. Greenwood)